Beyond Diet: "The Functional Nutrition BluePrint"

Optimizing Health from the Inside Out

Brooke

Nutrition Talk

Copyright © 2024 Shana Brooke

The moral right of the author has been asserted.

All rights reserved.
No part of this publication may be reproduced, stored in a retrieval system, or transmitted, in any form or by any means, without the prior permission in writing of the publisher, nor be otherwise circulated in any form of binding or cover other than that in which it is published and without a similar condition including this condition being imposed on the subsequent purchaser.

Published by Nutrition Talk

ISBN 979-8-3448-8321-2

Typesetting services by BOOKOW.COM

To Barbara Sanders my beautiful mother
(Inside and Out) January 11, 1949-2020

Preface

How I began on my Nutrition Journey...

I was in my early twenties and I did not have health insurance for the first time. I fell sick. The kind of sick that wipes your energy and all you can do is rest, hoping that it will make it better. I wasn't sure if it was the flu, a sinus infection, or just a virus. So I slept and I just felt worst. I didn't have the money to pay for a doc visit nor the meds that would have followed.

I went to a local HEALTH FOOD STORE and talked to the functional Practitoner. She advised me a couple of things do for to my symptoms at the time. I walked out of there with spending just $100. That was a lot but nearly not what it would have been.

I went back home and took what she advised and went to lay back down. In just a half an hour I was up and felt ninty percent better.

That then sent me on my journey. I stared down on a rabbit whole of research on herbal nutrition and by my late twenties I became a Certified Health Coach. I did that for more knowledge and to help myself and my loved ones. Now here in my fouties Im writing to you Spreading the knowledge and most importantly the truth. God's Truth.

Contents

Introduction to Functional Nutrition   1

Chapter 1: Building a Foundation for Health   6

Chapter 2: Personized Nutrition   11

Chapter 3: Key Nutrition Facts for a Healthy Life   15

    Chapter 4: Supporting Health with Supplements 26

Chapter 5: Functional Foods and Their Benefits   33

    Chapter 6: Meal Planning and Lifestyle Tips . 35

Conclusion: Your Path to Lasting Health   43

Introduction to Functional Nutrition

Whole Foods First Approach

Natural Foundation: Whole foods are the cornerstone of a healthy diet. They're rich in vitamins, minerals, fiber, and phytonutrients that our bodies recognize and can efficiently use. By prioritizing whole foods, we support the body's natural balance and function.

Bioavailability & Balance:

Nutrients from whole foods are often more bioavailable because they come in a form that includes co-factors that aid absorption and effectiveness. For example, an orange provides not just vitamin C but also fiber, flavonoids, and antioxidants that work synergistically.

Supporting Local and Seasonal Options: Whenever possible, our store aims to provide information and resources for locally sourced, seasonal produce to promote freshness, sustainability, and nutrient density. Eating seasonally can help customers align their diet with what their body naturally craves during different times of the year.

Quality Supplements as a Bridge, not a Replacement

Filling Nutritional Gaps

We recognize that in today's world, it's not always possible to meet every nutrient need through food alone. Factors like soil depletion, modern food processing, and hectic lifestyles can impact nutrient intake. Quality supplements offer a way to fill these gaps without compromising health.

Strict Ingredient Standards: We only offer supplements made from the purest ingredients, with minimal fillers, binders, and artificial additives. We seek products that are backed by rigorous testing, ethical sourcing, and transparent manufacturing practices.

Functional Formulas

When selecting supplements, we focus on bioavailability and functional formulas that provide benefits beyond just meeting nutrient needs. For example, instead of isolated vitamins, we might recommend a blend that includes supporting compounds or co-factors for optimal absorption.

Education and Empowerment Empowering Through

Knowledge: Our goal is to guide customers in making informed choices. We offer resources, one-on-one consultations, and access to insights from functional practitioners so customers can understand the "why" behind our recommendations.

Individualized Wellness: We recognize that each person's needs are unique. Whether someone's goal is energy, immunity, or digestive health, we're committed to providing customized guidance that respects their personal journey.

Commitment to Transparency

Label Honesty and Testing: We prioritize brands that disclose all ingredients and undergo third-party testing. Customers deserve to know exactly what they're putting into their bodies.

Responsibility for Sustainability: We also believe in supporting brands that care for the planet as much as our customers' health, with eco-friendly practices, sustainable sourcing, and reduced packaging where possible.

What to expect when collaborating with a Functional Practitioner

Personalized Guidance:

Functional practitioners often focus on root causes and holistic wellness, which aligns well with nutrition. They can guide customers on individualized supplement recommendations that complement their unique health needs, increasing customer trust and satisfaction.

Educational Opportunities: Partnering on workshops or informational sessions can provide value to customers and establish your store as a trusted source of knowledge in wellness. Topics could include gut health, detoxification, hormone balance, or immune support—common areas of functional health.

Product Recommendations: Practitioners can help customers navigate your product offerings based on their current health goals. This targeted approach can improve the effectiveness of the products for customers, potentially leading to higher sales and repeat business.

Referral Pathways: As customers visit your store seeking support, those with complex health concerns could be referred to the practitioner for more specialized guidance. Similarly, the practitioner may refer clients to your store for recommended products, boosting traffic.

Building Credibility: Having an expert endorsement from a functional practitioner can add credibility to your store. This collaboration can demonstrate your commitment to evidence-based, functional health approaches, appealing to customers looking for trustworthy guidance.

Chapter 1: Building a Foundation for Health

The key principles of nutrition are foundational for supporting health and wellness, and they generally emphasize balance, variety, and moderation. Here's an overview of the main principles:

Adequacy

Ensure sufficient intake of essential nutrients (carbohydrates, proteins, fats, vitamins, minerals, and water) to meet the body's needs. Adequate nutrition supports cellular functions, energy production, and immunity.

Balance

Incorporate a variety of nutrient-dense foods from all food groups (fruits, vegetables, grains, proteins, and dairy or dairy alternatives) to create balanced meals.

This helps avoid overemphasizing certain nutrients while missing out on others.

Variety

Consume a wide range of foods to ensure a comprehensive nutrient profile. Different foods provide unique nutrient profiles, so eating a diverse diet helps to cover all nutritional bases.

Moderation

Focus on portion control and avoid excessive intake of certain nutrients, especially saturated fats, sugars, and sodium. Moderation also applies to calorie intake, ensuring it aligns with the body's energy needs.

Nutrient Density

Opt for foods high in vitamins, minerals, and other beneficial compounds relative to their calorie content. Nutrient-dense foods like vegetables, fruits, lean proteins, and whole grains provide essential nutrients without excessive calories.

Energy Balance

Balance calorie intake with energy expenditure. Energy needs vary based on age, gender, body size, and activity level, so maintaining this balance helps manage body weight and avoid metabolic issues.

Personalization

Tailor nutrition to individual needs, considering factors like age, health status, physical activity, and personal preferences. Customized plans, especially those based on individual goals or medical needs, can be more effective.

Sustainability

Choose foods that support both individual health and environmental health. This may involve focusing on plant-based foods, reducing food waste, and selecting sustainable sources.

These principles form the basis of a nutritious diet that promotes health, prevents chronic disease, and supports energy and vitality for daily life.

Whole foods, nutrition density, and balance form the foundation of a truly nourishing diet, as the forms provide the body with essential nutrients in the most bioavailable while promoting overall health.

Whole Foods: Whole foods, such as fresh fruits, vegetables, whole grains, and minimally processed proteins, offer nutrients in their most natural state. Unlike highly processed foods, which often lack essential vitamins, minerals, and fiber, whole foods contain beneficial compounds like antioxidants and phytonutrients that support immunity, reduce inflammation, and improve digestion.

Nutrient Density:

Nutrient-dense foods provide a high number of vitamins, minerals, and other health-promoting components relative to their caloric content. Foods like leafy greens, berries, nuts, seeds, and lean proteins are nutrient-dense, meaning they supply essential nutrients without excess calories. Prioritizing nutrient dense foods helps prevent nutrient deficiencies, improves energy levels, and supports bodily functions.

Balance

Balanced nutrition involves the proper ratio of macronutrients (carbohydrates, proteins, and fats) and micronutrients (vitamins and minerals).

Achieving balance supports hormone regulation, energy production, and mental clarity. Additionally, a balanced diet with a variety of foods supports the gut microbiome, which is essential for digestive health and immunity.

Together, whole foods, nutrient density, and balance provide a powerful approach to health that goes beyond calorie counting, focusing instead on nourishing the body fully. This approach can help prevent chronic diseases, improve mood, and enhance overall quality of life.

Functional nutrition takes a holistic and individualized approach to diet, focusing on whole foods and considering how various factors, such as **lifestyle, genetics, environment, and unique health conditions, impact a person's nutritional needs**. **Unlike traditional diets** that often aim for weight loss or general health by following standardized guidelines, functional nutrition looks at how foods impact specific bodily systems to promote optimal health.

In functional nutrition, the emphasis is on nutrient-dense foods, **identifying and addressing any underlying imbalances (like inflammation or gut health issues), and adjusting the diet to support healing and long-term wellness**. This approach often involves regular testing and close tracking of symptoms to fine-tune nutritional strategies, making it highly personalized.

Chapter 2: Personized Nutrition

Understanding Bio individuality

Bio-individuality is the concept that each person has unique nutritional and lifestyle needs based on their genetics, environment, lifestyle, and even microbiome. It acknowledges that there is no one-size-fits-all approach to health and wellness—what works for one person may not work for another. This concept is often used in functional medicine, where practitioners consider individual differences when designing dietary, supplement, and lifestyle recommendations.

Here are some aspects that influence bio individuality>

Genetics: Our genes influence how we metabolize nutrients, process toxins, and react to specific foods.

For example, some people are more sensitive to caffeine, while others can handle dairy better than others.

Microbiome: The bacteria in our gut influence digestion, nutrient absorption, and even mood. Since everyone has a unique microbiome, dietary needs vary.

Lifestyle: Stress, sleep, exercise, and environment all shape our health needs. For example, someone with a high-stress lifestyle may need more magnesium or adaptogenic herbs.

Health History: Past health conditions, medications, and surgeries can impact nutritional needs.

Metabolism and Hormones: Metabolic rate and hormone levels differ, impacting dietary needs and exercise requirements.

Understanding bio-individuality helps you make informed choices in tailoring nutrition and wellness practices to a person's specific needs, which is especially relevant in a personalized or functional approach to health.

Listening to the body is essential because each person has unique needs based on factors like

lifestyle, genetics, environment, and

current health status. When we tune in to signals like energy levels, sleep quality, mood, digestion, and cravings, we can better understand what our bodies are truly asking for, rather than following generic advice that may not apply to us. For instance, one person may thrive on a high-protein diet, while another might feel better with more plant-based foods or specific supplements.

By adjusting according to these unique cues, we can improve our resilience, reduce stress, and make sustainable progress. Working with a functional practitioner can help identify underlying issues, guide adjustments, and tailor interventions to suit individual needs, especially when it comes to personalized nutrition, exercise, and lifestyle practices. This individualized approach promotes lasting wellness by fostering balance and addressing root causes, rather than just treating symptoms.

A food monitoring journal can be a fantastic ...

tool for identifying patterns in your diet, energy, digestion, and mood. Here's a simple template you might find helpful:

Date and Time: Note when you ate, as timing can impact digestion, energy, and sleep.

Meal and Ingredients: Describe the meal and ingredients (including portion sizes, if possible).

Hunger Level (1-10): Rate hunger before eating to understand if you're eating out of habit, stress, or actual hunger.

Mood & Emotions: Track how you felt before and after eating. Food can impact mood, and emotions can impact food choices.

Digestion & Physical Symptoms: Note any physical reactions (like bloating, gas, or energy levels).

Cravings: Document any cravings, as these can highlight nutrient deficiencies or emotional triggers.

Hydration & Activity: Include water intake and exercise to see how they interact with your diet.

Apps like MyFitnessPal, Cronometer,

Blood Type Buddy or even a custom notebook can be a great way to keep track. Working with your functional practitioner on this can bring insights into areas needing adjustment.

Chapter 3: Key Nutrition Facts for a Healthy Life

Here's a quick guide on essential vitamins, minerals, and micronutrients, their benefits, and sources:

Vitamins

<u>Vitamin A</u>

Benefits: Supports vision, immune health, and skin health and LIVER

Sources: Carrots, sweet potatoes, spinach, and liver.

<u>Vitamin C</u>

Benefits: Boosts immunity,

promotes skin health, and aids iron absorption.

Sources: Oranges,

strawberries, bell peppers, and broccoli, dark leafy greens

Vitamin D

Benefits: Vital for bone health and immune support.

Sources:

Sunlight, fortified dairy, egg yolks, and fatty fishlike salmon.

Vitamin E

Benefits:

Acts as an antioxidant, supports skin health, and helps immune function.

Sources

Nuts, seeds, spinach, and sunflower oil.

Vitamin K

Benefits: Essential for blood clotting and bone health.

Sources: Leafy greens

(kale, spinach), broccoli, and Brussels sprouts.

B Vitamins (B1, B2, B3, B5, B6, B7, B9, B12)

Benefits:

Important for energy production, brain function, and red blood cell formation.

Sources:

Whole grains, eggs, dairy, meat, legumes, and leafy greens.

Minerals

Calcium

Benefits: Critical for bone health, nerve function, and muscle contraction.

Sources

Dairy, fortified plant milks, leafy greens, and almonds.

Iron

Benefits: Key for hemoglobin production and oxygen transport in blood.

Sources: Red meat, beans,

lentils, spinach, and fortified cereals.

Magnesium

Benefits:

Supports muscle and nerve function, blood pressure regulation, and bone health.

Sources: Nuts, seeds, leafy greens, and whole grains.

Potassium

Benefits: Essential for

heart health, fluid balance, and muscle function.

Sources:

Bananas, potatoes, spinach, and avocados.

Zinc

Benefits:

Supports immune function, wound healing, and DNA synthesis.

Sources: Meat, shellfish, legumes, red onions, and seeds

Sodium

Benefits: Helps with fluid

balance, nerve signals, and muscle contractions.

Sources: Table salt, processed foods, and natural sources like celery.

Key Micronutrients

Iodine

Benefits: Essential for thyroid health and metabolic regulation.

Sources:

Iodized salt, seafood, dairy, and eggs.

Selenium

Benefits:

Acts as an antioxidant, supporting immune health and thyroid function.

Sources

Brazil nuts, seafood, and organ meats.

Copper

Benefits:

Important for iron metabolism, brain function, and immune health.

Sources:

Shellfish, nuts, seeds, and whole grains.

Manganese

Benefits:

Supports bone health, metabolism, and antioxidant functions.

Sources:

Whole grains, nuts, leafy greens, and tea.

Chromium

Benefits: Helps regulate blood sugar by enhancing insulin action.

Sources: Broccoli, potatoes, whole grains, and lean meats.

Tips for Maximizing Absorption

Pairing: Some nutrients, like iron, are better absorbed with vitamin C

Fat-Soluble Vitamins: Vitamins A, D, E, and K are absorbed with dietary fat.

Avoiding Excessive Processing: Cooking can sometimes reduce nutrient levels, especially water-soluble vitamins (B and C).

There are plenty of other pairings that are important as well. Ask your Functional Practitioner

Essential vitamins, nutrients, minerals, and micronutrients play crucial roles in maintaining overall health and supporting various bodily functions. Here's a brief overview of how they work in the body:

Vitamins

Function: Vitamins are

organic compounds that are essential for metabolic processes. They are categorized into water-soluble (e.g., B vitamins, vitamin C) and fat-soluble (e.g., vitamins A, D, E, K).

Role:

Energy Production:

B Vitamins are vital for energy metabolism.

Antioxidant Activity:

Vitamins C and E help protect cells from oxidative stress.

Bone Health: Vitamin D aids calcium absorption, essential for bone health.

Immune Function:

Vitamins A and C support immune responses.

Minerals

Function: Minerals are inorganic elements that support a variety of functions in the body, such as building bones and teeth, aiding nerve function, and muscle contraction.

Role:

Bone Structure:

Calcium and phosphorus are crucial for bone health.

Electrolyte Balance: Sodium, potassium, and chloride help maintain fluid balance and nerve function.

Hemoglobin Formation:

Iron is essential for the formation of hemoglobin, which carries oxygen in the blood.

Macronutrients

Carbohydrates: Provide energy for the body. They are broken down into glucose, which is used for immediate energy or stored as glycogen.

Proteins: Made of amino acids, proteins are essential for tissue repair, muscle growth, and enzyme production.

Fats: Important for hormone production, cell membrane structure, and energy storage. Essential fatty acids (like omega-3s) are crucial for brain health.

Function: Micronutrients, which include vitamins and minerals, are needed in smaller amounts but are vital for numerous biochemical processes.

Role:

Cofactors: Many vitamins and minerals act as cofactors for enzymes, facilitating biochemical reactions.

Regulation: They help regulate processes like gene expression, cell signaling, and hormone production.

Hydration

- **Role:** While not a vitamin or mineral, water is crucial for transporting nutrients, regulating body temperature, and maintaining cellular functions.

Summary

Essential vitamins, nutrients, minerals, and micronutrients work synergistically to support metabolism, maintain cellular integrity, and ensure overall health.

A balanced diet rich in these components is essential for preventing deficiencies and promoting optimal body function.

Here are some practical tips for reading nutrition labels and making informed decisions:

Understanding Nutrition Labels

Start with Serving Size: Check the serving size at the top of the label. All nutritional information is based on this amount, so adjust based on how much you consume.

Look at Calories: Consider your daily calorie needs and how the product fits into your overall diet. Pay attention to both calories and calories from fat.

Focus on Nutrients:

Limit: Saturated fat, trans fat, cholesterol, and sodium. Aim for low amounts of these in your diet.

Get Enough: Dietary fiber, vitamins (like A, C, and D), and minerals (like calcium and iron). These are beneficial for your health.

Watch for Added Sugars: Added sugars can contribute to weight gain and health issues. Look for products with minimal or no added sugars.

Check the Ingredients List: Ingredients are listed in order by weight. The first few ingredients make up most of the product, so look for whole foods and avoid products with many artificial ingredients or preservatives.

Making Informed Decisions

Use the 5-20 Rule: For nutrients to limit (like saturated fat and sodium), aim for 5% or less of the Daily Value (DV). For nutrients to get enough of (like fiber), look for 20% or more of the DV.

Beware of Health Claims: Terms like "organic," "low-fat," or "sugar free" can be misleading. Always check the label for actual nutrient content.

Portion Control: Be mindful of portion sizes, especially with snacks and high calorie foods. It's easy to overconsume if you're not careful.

Plan Your Meals: Take time to plan meals around whole, nutrient dense foods. This can help you avoid processed foods and make healthier choices.

Educate Yourself: Keep learning about nutrition and how different foods affect your body. Resources like books, reputable websites, or workshops can provide valuable information.

Additional Tips:

Stay Hydrated: Don't forget about hydration. Water is essential for overall health.

Consider Whole Foods: Focus on whole, unprocessed foods, which are typically more nutritious and beneficial.

Mindful Eating: Pay attention to your hunger cues and eat slowly. This can help you enjoy your food more and avoid overeating.

By applying these tips, you can become more confident in your ability to read labels and make healthier food choices.

Chapter 4: Supporting Health with Supplements

Supporting health with supplements can be beneficial when used thoughtfully and in combination with a balanced diet and lifestyle. Here's a general guide:

Foundation:

Multivitamins

- A well-formulated multivitamin can cover basic needs, especially if diet gaps exist. Look for products with natural forms of vitamins, like methylated B vitamins, and avoid synthetic additives.

Gut Health: Probiotics & Digestive Enzymes

Probiotics: Support the gut microbiome, which plays a role in digestion, immunity, and mental health. Look for strains like *Lactobacillus* and *Bifidobacterium*.

Digestive Enzymes: Helpful for those who experience bloating, especially after meals. These assist in breaking down carbs, fats, and proteins more efficiently.

Essential Fatty Acids: Omega-3s

- Omega-3 fatty acids (from fish oil or algae) support heart, brain, and joint health. EPA and DHA are particularly beneficial, and a dose around 1,000-2,000 mg per day is common.

Bone & Muscle Health:

Calcium, Magnesium, and Vitamin D

Vitamin D3: Vital for immune health and calcium absorption. Many people have low levels, so supplementing with 1,000-5,000 IU can be beneficial, especially in winter.

Magnesium:

Supports muscle relaxation, sleep, and bone health. Forms like magnesium glycinate or citrate are often better absorbed.

Calcium: Important if dietary intake is low, especially for bone density.

Antioxidants: Vitamin C, E, and CoQ10

Vitamin C and E: Powerful antioxidants that support immune health, skin, and overall cellular protection.

CoQ10: Beneficial for heart health and energy production, especially if on statin medications.

Energy & Stress Support:

B Vitamins and Adaptogens

B-Complex:

Supports energy production and stress resilience. Look for a complete B complex with methylated B12 and folate for better absorption.

Adaptogens (e.g., Ashwagandha, Rhodiola): Help the body adapt to stress and support mental clarity and energy.

Sleep & Relaxation: Melatonin, Magnesium, and Herbal Options

Melatonin: A small dose (0.5-3 mg) can help regulate sleep cycles, particularly if your routine is disrupted.

Herbs (e.g., Valerian, Chamomile): Natural calming agents that promote relaxation without causing drowsiness the next day.

Customized Supplementation:

For more specific needs like hormone balance, cognitive support, or athletic performance, it's often best to consult a healthcare provider. A functional practitioner, like the one you're associated with, can guide you on tailored supplement strategies and testing.

Using supplements wisely can enhance a balanced diet, filling in nutritional gaps while supporting overall health. Here's how to

approach it:

Identify Your Needs.

Prioritize Food First.

Build a Strong Base with Whole Foods: Whole foods provide a wide range of nutrients and fiber, which supplements can't fully replicate.

Use Supplements to Fill Gaps: Supplements should target nutrients that are harder to obtain from food alone or in cases of confirmed deficiencies.

Choose Quality Supplements

Look for Reputable Brands: Consider brands that use high-quality ingredients and have transparent sourcing practices. Look for third-party testing, which ensures that what's on the label is in the product.

Avoid Fillers and Additives: Aim for supplements with minimal additives, artificial colors, or preservatives.

Dosage and Timing

Follow Recommended Dosages:

Avoid megadose unless advised by a healthcare provider, as more isn't always better and can sometimes be harmful.

Consider Timing and

Combinations: Some supplements are best taken with food (e.g., fat-soluble vitamins like D, E, K) or separately (e.g., calcium and iron compete for absorption).

Be Mindful of Interactions

Check for Interactions: Some supplements interact with medications or other nutrients. For instance, vitamin K can interfere with blood-thinning medications.

Monitor Side Effects: If you notice any adverse effects, stop and consult a healthcare provider.

Consistency and Regular Monitoring

Stick to a Routine: Supplements need consistency to show effects, so try to incorporate them into your daily routine.

Re-evaluate Periodically: Nutritional needs can change, so reassessing your supplement regimen annually or as needed can keep it aligned with your health goals.

Examples of Common

Supplements and Their Uses

Vitamin D: Supports bone health and immune function, especially for those with limited sun exposure.

Magnesium: Aids muscle function, sleep, and relaxation.

Omega-3 Fatty Acids: Beneficial for heart and brain health, especially if you don't consume enough fatty fish.

The key to wise supplementation is a targeted, informed approach that complements— not replaces —a healthy diet.

Chapter 5: Functional Foods and Their Benefits

Functional foods are foods that offer health benefits beyond basic nutrition, often containing bioactive compounds that support specific physiological functions. Here's a breakdown of key benefits and examples:

Heart Health

Example: Oats, rich in beta-glucan, can help lower LDL cholesterol.

Benefits: May reduce heart disease risk by improving lipid profiles and reducing blood pressure.

Digestive Health

Example: Probiotic-rich foods like yogurt and fermented products support gut health.

Benefits: Enhances digestion, supports gut microbiome balance, and can aid immune function.

Antioxidant and Anti-inflammatory Effects

Example: Berries, rich in flavonoids, and green tea are potent sources.

Benefits: Reduces cellular damage, potentially lowering cancer risk and supporting cognitive health.

Immune System Support

Example: Garlic and ginger have natural immune-boosting properties.

Benefits: Contains bioactive like allicin in garlic, which support the immune response.

Bone Health

Example: Fortified dairy products, like milk with added vitamin D.

Benefits: Enhances calcium absorption and supports bone density, reducing osteoporosis risk.

Metabolic Health and Blood Sugar Regulation

Example: Cinnamon and whole grains, which have a low glycemic impact.

Benefits: Improves blood sugar control, supporting metabolic health and reducing diabetes risk.

Functional foods are a popular topic in nutritional science and a focus in functional medicine for their preventive and therapeutic roles.

Chapter 6: Meal Planning and Lifestyle Tips

Here are some meal planning and lifestyle tips to support overall health,

Meal Planning Tips

Balance Macros: Focus on balanced macronutrients in each meal—protein, healthy fats, and complex carbs. This combination supports steady energy and blood sugar levels.

Prep Ahead

Spend a couple of hours each week prepping meals or components like chopped veggies, cooked grains, or proteins. This makes it easier to stick to your plan on busy days.

Prioritize Protein: Include a protein source in every meal (e.g., eggs, legumes, lean meats, or plant-based options) to support satiety and muscle repair.

Fiber-Rich Foods:

Aim for vegetables, whole grains, fruits, and legumes. Fiber supports gut health, slows digestion, and stabilizes blood sugar levels.

Healthy Snacks:

Keep nutritious snacks like nuts, yogurt, or fruit on hand. This can prevent energy crashes and overeating later.

Hydration: Incorporate hydrating foods like cucumbers, lettuce, and oranges. Drinking water throughout the day is key for digestion, energy, and concentration.

Lifestyle Tips

Sleep: Aim for 7-9 hours of quality sleep. Consistent, restful sleep is essential for energy, mental clarity, and immune support.

Exercise Regularly: Even short sessions can make a difference. Try to include both aerobic and strength-based activities for heart and muscle health.

Mindful Eating: Take time to enjoy your meals and eat without distractions. It aids digestion, allows you to better tune into hunger and fullness cues, and improves satisfaction.

Routine Building: A consistent routine helps in reducing stress, improving sleep, and promoting overall wellness. Try incorporating morning or evening rituals that relax and rejuvenate.

Limit Processed Foods: These are often high in added sugars, unhealthy fats, and preservatives, which can impact energy and health long-term. Opt for whole foods when possible.

Stay Connected: Regular social connections positively impact mental health. If possible, eat meals with friends or family, which can improve your mood and overall wellbeing.

Would you like tips specific to any health goals or areas?

Here are three sample meal plans based on functional nutrition principles, designed to support whole-body health, balance blood sugar, and reduce inflammation. These can be tailored to specific health needs, dietary restrictions, and preferences.

Meal Plan 1: Energy and Blood Sugar Balance

Breakfast

Green smoothie: Spinach, kale, cucumber, avocado, half an apple, and a tablespoon of ground flaxseed, blended with unsweetened almond milk.

Snack

A handful of **almonds and blueberries** for antioxidants and healthy fats.

Lunch

Chickpea and vegetable salad: Mixed greens with chickpeas, diced cucumber, cherry tomatoes, radishes, and a sprinkle of pumpkin seeds.

Dress with extra virgin olive oil and apple cider vinegar for gut health.

Snack

Celery sticks with almond butter and a sprinkle of cinnamon.

Dinner

Turmeric spiced lentil soup: red lentils, turmeric, cumin, carrots, spinach, and onion cooked in vegetable broth.

Serve with a small side of fermented veggies, like sauerkraut, for added probiotics.

Meal Plan 2: Hormone Balance and Detox Support

Breakfast

Chia pudding made with coconut milk and topped with berries, a few almonds, and a sprinkle of cacao nibs for magnesium.

Snack

Sliced apple with tahini or almond butter for healthy fats.

Lunch

Broccoli and kale salad with organic, free-range chicken, topped with avocado, beets, and sunflower seeds.

Dress in lemon-tahini dressing for liver support.

Snack

Pumpkin seeds and a few Brazil nuts for selenium and zinc, important for thyroid health.

Dinner

Wild-caught cod baked with fresh herbs, served with roasted sweet potatoes and steamed asparagus.

Drizzle with olive oil for healthy fats and add a pinch of sea salt for minerals.

Notes:

Hydration: Drink water throughout the day, optionally adding lemon or herbal teas.

Herbs and Spices:

Incorporate anti-inflammatory options like turmeric, ginger, and cinnamon for their functional benefits.

Preparation Tips

Batch cook grains, proteins, and roasted veggies at the start of the week to simplify meal assembly.

Creating a sustainable eating routine revolves around balance, consistency, and flexibility. Here are some practical tips to help:

Prioritize Whole Foods: Aim for meals centered around whole foods like vegetables, fruits, lean proteins, whole grains, and healthy fats. Minimally processed foods provide more nutrients and can be more satisfying.

Plan Ahead but Stay Flexible: Meal planning is helpful but adaptable. Prep ingredients or meals for the week to avoid last-minute unhealthy choices but allow yourself the flexibility to switch things up.

Start with Small Changes:

Instead of a full overhaul, introduce one or two small adjustments each week, like adding more greens to meals or cutting back on sugar. Small changes can become sustainable habits over time.

Focus on balance, Not Restriction: Include a variety of foods without overly restricting any food group. Having a good balance keeps you nourished and helps reduce cravings.

Listen to Your Body: Pay attention to hunger and fullness cues to avoid overeating or undereating. Mindful eating practices, like eating slowly and without distractions, can help with this.

Stay Hydrated: Drinking water throughout the day aids digestion and helps you feel more satisfied. Sometimes hunger is mistaken for thirst, so stay hydrated to avoid unnecessary snacking.

Prioritize Protein and Fiber: Protein and fiber help keep you full for longer and stabilize blood sugar levels. Include a source of each at every meal, like eggs and greens for breakfast or beans and veggies for lunch.

Be Kind to Yourself: Building a sustainable routine takes time, and it's normal to have slip-ups. Focus on progress rather than perfection, and don't let small setbacks derail your efforts.

Prepare for Social

Situations: Plan for events by eating balanced meals during the day. At the event, enjoy yourself without feeling the need to restrict or overindulge.

Evaluate and Adjust Regularly: Every few weeks, review what's working and what isn't. This will help you make minor adjustments and keep the routine enjoyable and aligned with your goals.

Conclusion: Your Path to Lasting Health

Creating a path to lasting health is about building a sustainable lifestyle that nurtures both body and mind. It involves understanding your unique needs, setting realistic goals, and making consistent, positive changes.

Here's a summary of the journey:

Awareness: Recognize your current health state and identify areas for improvement. This often begins with assessments, such as those from a functional practitioner, to understand your body's specific needs.

Education: Equip yourself with knowledge about nutrition, exercise, sleep, stress management, and other aspects of health. Learning about how food and lifestyle impact your body empowers you to make informed choices.

Individualized Nutrition: Fueling your body with foods that support energy, mental clarity, and physical resilience. Using your nutrition store's resources and expertise, you can ensure access to high-quality supplements or specialty foods that meet unique dietary needs.

Exercise and Movement:

Incorporate consistent, enjoyable physical activity into your routine. Regular movement boosts energy supports a healthy metabolism and promotes mental well-being.

Mind-Body Connection: Practice mindfulness, manage stress, and prioritize sleep. Health is as much mental and emotional as it is physical.

Consistency: Sustainable health doesn't happen overnight. Small, consistent actions build a foundation that, over time, leads to significant results.

Support and Accountability: Work with a support network, like a functional practitioner, friends, or a coach, to help keep you on track and address any challenges that arise.

By following these principles, you build a holistic and personalized path to lasting health—one that aligns with your goals and is sustainable for the long term.

Remember, the journey is ongoing, with each day an opportunity to reinforce your commitment to health and well-being.

Making small, consistent changes is one of the most effective ways to achieve long-lasting results. Often, when we aim for big changes all at once, the effort can feel overwhelming, and we might lose momentum. But when you focus on tiny, manageable adjustments, you build habits over time, making it easier to stay on track and see real progress.

Start by choosing one small action— whether it's drinking an extra glass of water, walking for 10 minutes each day, or swapping out one processed snack for something healthier. It may feel minor, but the consistency of these small choice's compounds, creating a snowball effect that leads to significant, sustainable change. Remember, it's these steady, small steps that often take you the farthest.

Nourishing the body is essential for lifelong wellness, as it provides the foundation for physical and mental resilience. Our bodies are constantly renewing, regenerating, and adapting, and quality nutrition supplies the building blocks needed for these processes. A well-nourished body can better ward off disease, recover faster from illness or injury, and maintain energy and vitality.

Investing in a balanced, nutrient-rich diet not only enhances the body's natural defenses but also supports mental clarity, mood stability, and overall emotional well-being. Proper nourishment allows us to perform at our best, both physically and mentally, well into old age. It's more than just preventing deficiency; it's about thriving. Over time, consistent nourishment becomes a key to a longer, healthier, and more fulfilling life.

Integrating wellness habits like regular hydration, mindful eating, and supplementing where necessary is a proactive approach to lifelong wellness— investing in your health now can yield invaluable returns over the years.

Made in the USA
Columbia, SC
17 November 2024